MATH PUZZLES FOR KIDS

By

Adriana P. Jenova

Published by PUBLISHING COMPANY in 2017
First edition: First printing
Illustrations and design © 2017 Adriana P. Jenova

www.allcoloringbook.com

ISBN-13: 978-1542883023
ISBN-10: 1542883024

HOW MANY MONKEYS DO YOU SEE?

?

HOW MANY DOGS DO YOU SEE? ?

ANSWER 19

HOW MANY BIRDS DO YOU SEE? ?

HOW MANY MICE DO YOU SEE? ?

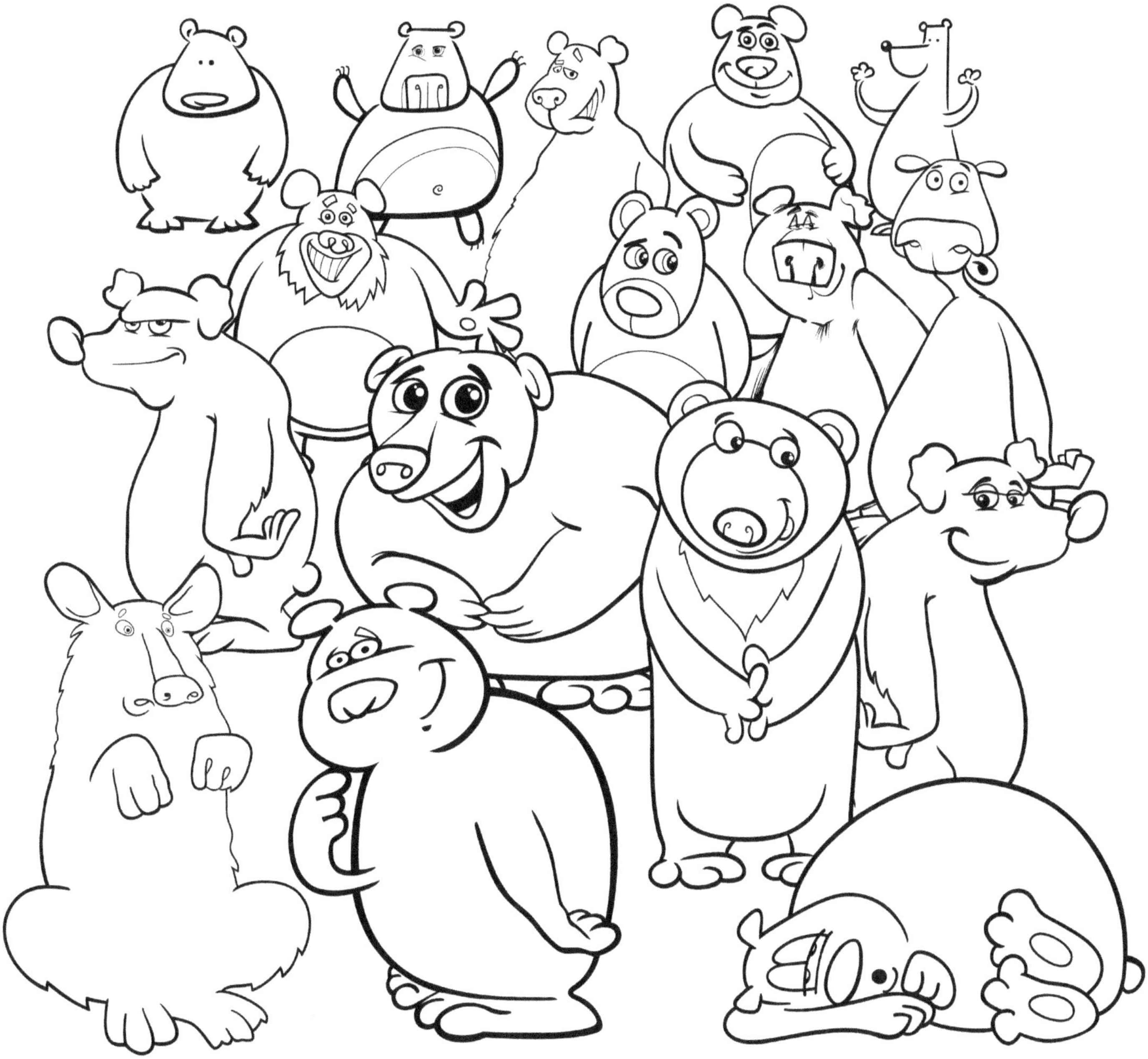

HOW MANY **BEARS** DO YOU SEE? ?

ANSWER 16

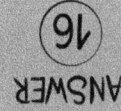

HOW MANY COWS DO YOU SEE? ?

HOW MANY ELEPHANTS DO YOU SEE? ?

ANSWER 10

HOW MANY ALIENS DO YOU SEE?

DO YOU SEE? ?

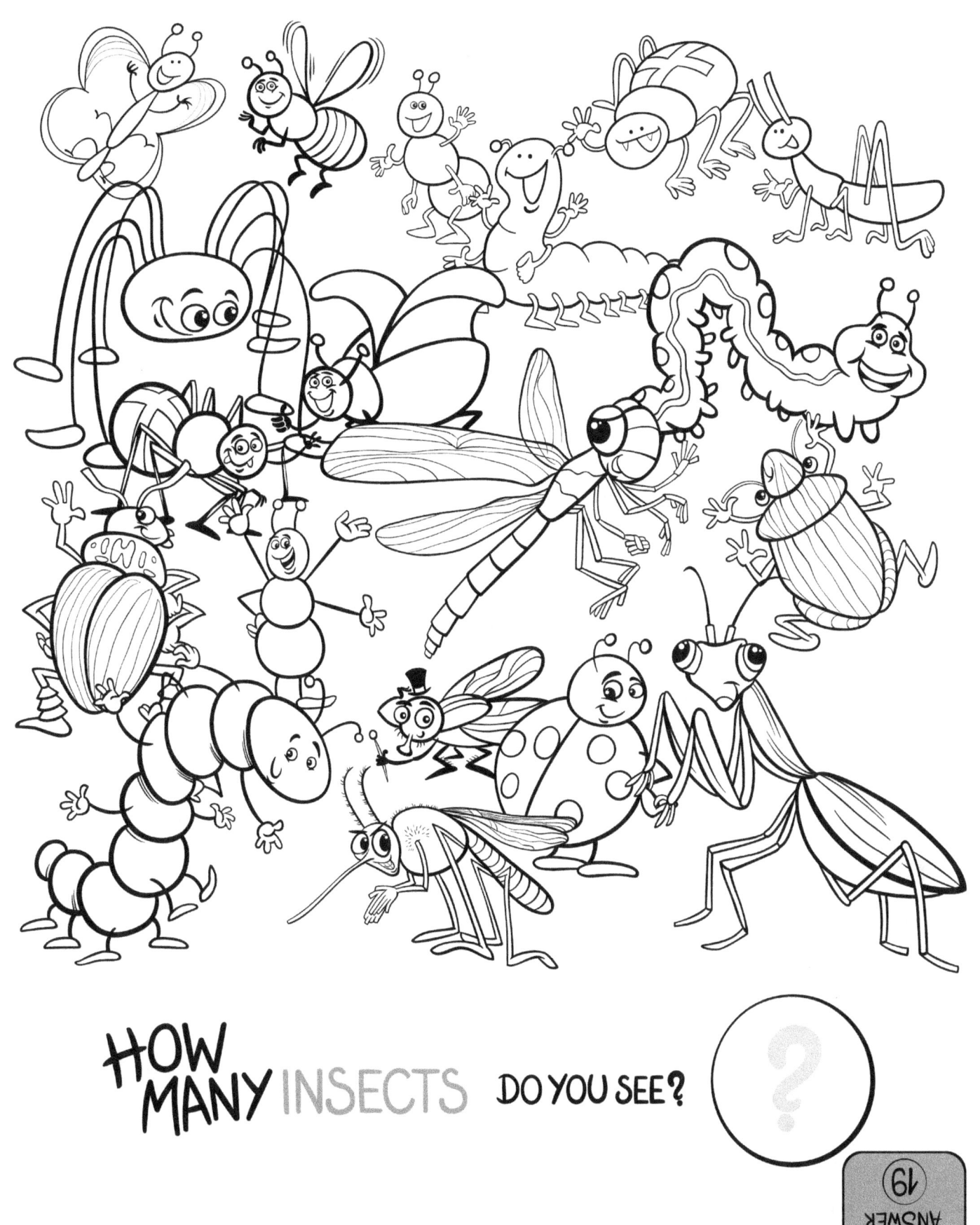

HOW MANY INSECTS DO YOU SEE? ?

HOW MANY SANTAS DO YOU SEE?

?

ANSWER 17

HOW MANY DINOS DO YOU SEE? ?

ANSWER 8

HOW MANY DRAGONS DO YOU SEE? ?

ANSWER 11

HOW MANY HORSES DO YOU SEE? ?

HOW MANY DONKEYS DO YOU SEE?

?

ANSWER 7

HOW MANY GOATS DO YOU SEE? ?

ANSWER 8

HOW MANY FISH DO YOU SEE? ?

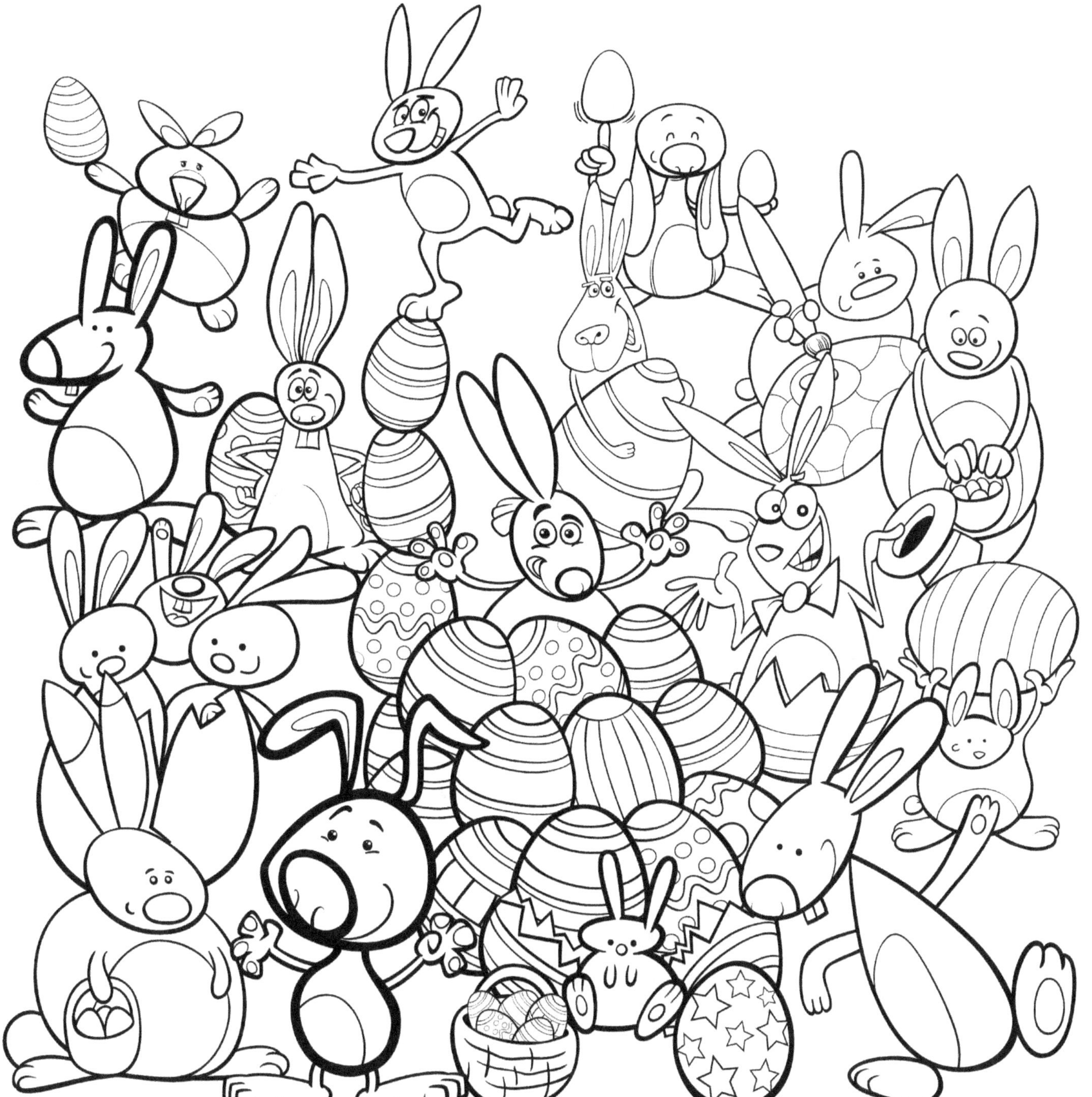

HOW MANY EASTER BUNNIES DO YOU SEE?

ANSWER 18

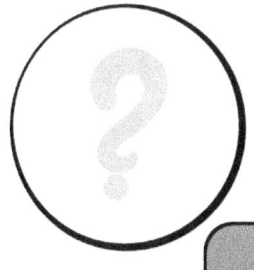

HOW MANY KIDS DO YOU SEE?

?

ANSWER 12

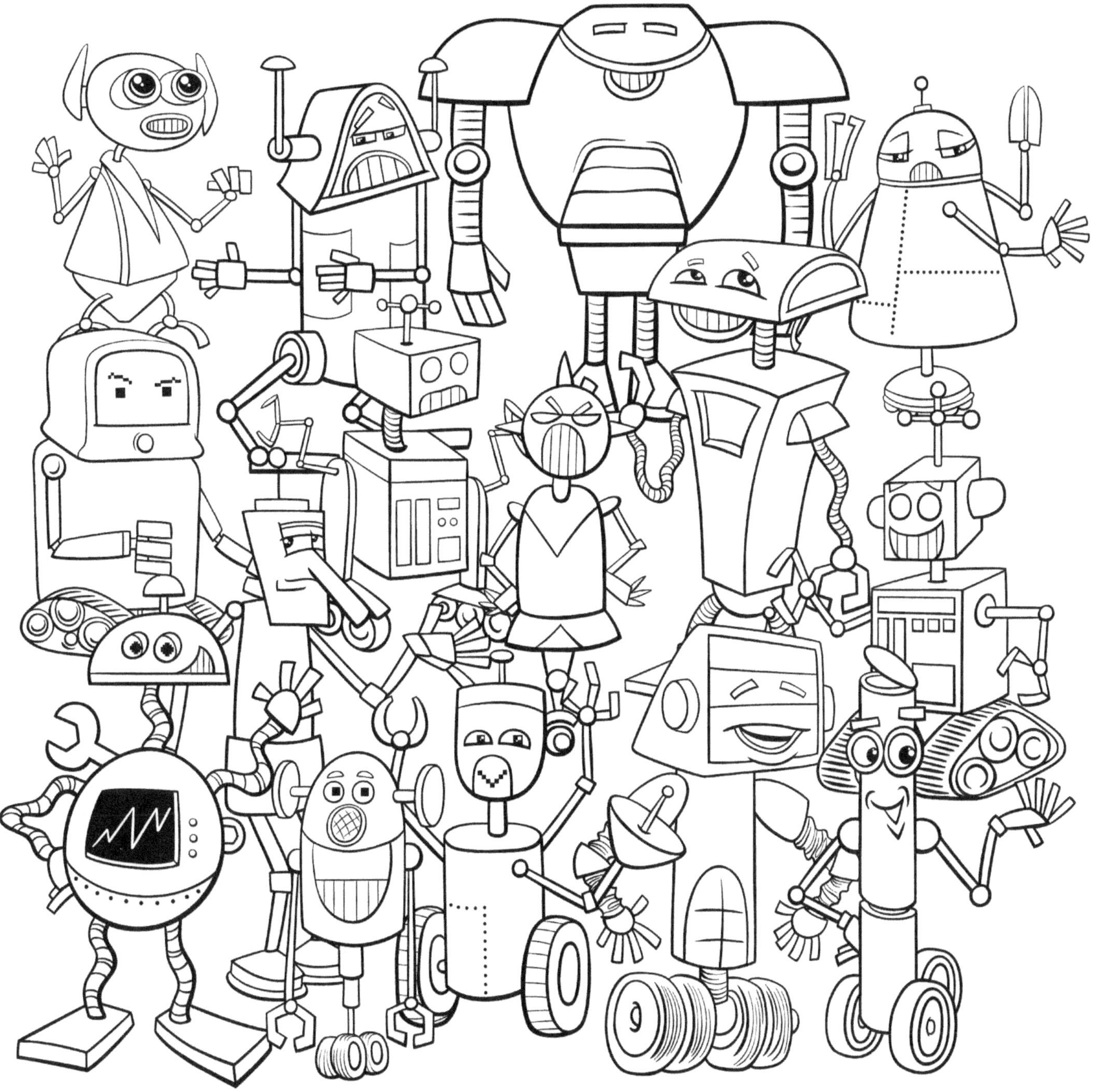

HOW MANY ROBOTS DO YOU SEE?

?

HOW MANY DOGS DO YOU SEE?

HOW MANY LEPRECHAUNS DO YOU SEE? ?

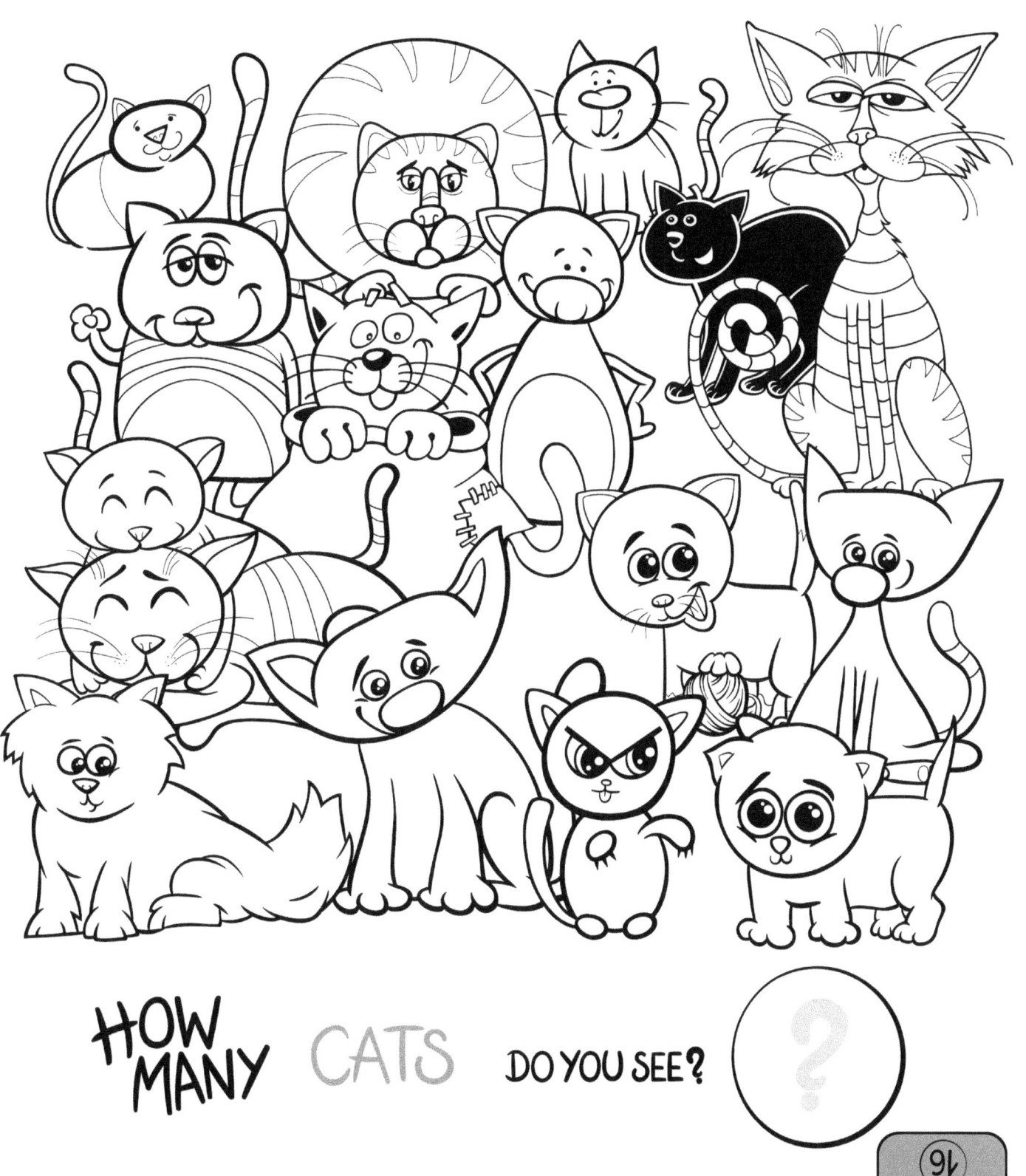

HOW MANY CATS DO YOU SEE?

?

HOW MANY ANIMALS DO YOU SEE?

?

ANSWER 13

Thank You

Hope you've enjoyed your reading experience.

We here at Adriana P. Jenova will always strive to deliver to you the highest quality guides.

So I'd like to thank you for supporting us and reading until the very end.

Before you go, would you mind leaving us a review on Amazon?

It will mean a lot to us and support us creating high quality guides for you in the future.

Thanks once again and here's where you can leave a review.

Get Free Illustration Coloring Page Below

https://www.facebook.com/Adrianapjenovacoloringbook

Warmly yours,

Adriana P. Jenova Team